96% Dark Universe
- Lost and Found -

Henryk Frystacki

96 % Dark Universe
- Lost and Found -

"From physics evolution…
…to view point revolution"

This review of classical space-time enlightens on hidden 96% of energies in our universe, just with few complementary and logical views on time and on space…

ISBN 978-1-4461-2806-0

© 2010 Dr. Henryk Frystacki
flucht.der.zeit@googlemail.com
**- Author's Edition of September 2010 -
Extract "Escape of Time" submitted to scientific publication and discussions.**

Contents

I) Introduction — 5

II) Complementary views on space-time — 7

III) Discussion of space-time leaps and conclusions — 13

IV) Definitions for the construction of a relativistic time suite — 16

V) Gravitational time dilation within the super symmetry — 33

VI) Reflections upon electron mass and atomic mass unit — 35

VII) Expedition through the standard model of physics — 37

References — 44

I) Introduction

The special theory of relativity[1] "STR" and the general theory of relativity[2] "GTR" introduce speed of light scaled time as fourth dimension, in addition to our three perceived space dimensions length, width, height. Euclidean geometry has been replaced by differential geometry. The GTR turns into the STR in case of sufficiently small areas in space-time, or an ideal, mass-free universe.

STR and GTR are based on the assumption and on many experimental confirmations of the constancy of speed of light in a vacuum[3], observed independently of any own motion or any other type of energetic influence. The consequences of this constancy of speed of light are time dilations, length contractions, and a relativity of simultaneity of events. STR describes additionally an energy equivalence of baryonic masses with $E=mc^2$. GTR explains gravitation as seeming force, caused by curvatures of space-time. The constancy of the speed of light in a vacuum was initially a postulate, stemmed from the assumption that the speed of light in Maxwell's equations of electromagnetism[4] stays constant in any inertial frame of reference, assuming homogeneous time and also homogeneous and isotropic space. Minkowski replaced Einstein's original concept of separated space and time by the construction of "space-time" and reformulated Einstein's work[5]. The paths of light in Minkowski's space-time-graph cover mathematically a zero space-time interval, having the impact that any observer will read the same constant value for the speed of light.

Asking Einstein about the nature of time, he answered that "time is what I can read on a clock", expressing that time describes the sequence of events, moving on in the present, with an origin in the past, and heading towards the future. The present of an observation of any event can be only defined in one single point in space-time. Other points neither placed in the past nor in the future are separated in space. Keeping the speed of light constant, space-time can be described with curvatures, as done by Einstein with energy tensors and relativistic field equations, explaining a far-reaching gravitation despite the restriction for information and energy transport at the speed of light. Einstein's cosmological constant[6], introduced and later dismissed by him, was re-introduced to adapt the general theory of relativity to the latest discoveries and current models of astrophysics[7]. However, STR and GTR have no proper explanation for the uncertainty principle of Heisenberg[8], stating that certain pairs of physical properties like position and momentum cannot be known to arbitrary precision. At known types of energy quantum levels, both are inconsistent with quantum mechanics.

Two equations of relativistic mechanics describe time dilation and length contraction of the special theory of relativity STR with the coordinates x', y', z' and t' of a moving system S' at the speed v, and with the coordinates x, y, z and t of the initial starting system S. Δx and Δx' are lengths, Δt and Δt' are time intervals read on clocks and compared with each other. System S' moves along the x-axis. c stands for the speed of light in a vacuum.

$$x = \frac{x' + vt}{\sqrt{1 - v^2/c^2}} \qquad t = \frac{(t' + \frac{v}{c^2}x')}{\sqrt{1 - v^2/c^2}}$$

Note: Time in non-coinciding x'-coordinates in S' differs, if evaluated in S. This describes a relativity of simultaneity of events. For two events in the same space location in S' ($x'_2 = x'_1$) but at a different time $t'_2 \neq t'_1$, time dilation of the special theory of relativity is described with $x'_2 - x'_1 = 0$ by:

$$\Delta t = \frac{\Delta t'}{\sqrt{1 - v^2/c^2}} + \frac{v}{c^2} \frac{x'_2 - x'_1}{\sqrt{1 - v^2/c^2}} = \frac{\Delta t'}{\sqrt{1 - v^2/c^2}} \geq \Delta t'$$

The time period of a moving observer is shorter in comparison with the time period of any remaining observer in the starting point: A moving observer stays continuously in the present of the remaining one but ages slower. The corresponding length contraction is based on proven constancy of speed of light, with the impact that the original distance Δx to a destination is getting shorter for the moving observer, expressed by Δx':

$$\Delta x = \frac{\Delta x'}{\sqrt{1 - v^2/c^2}} \geq \Delta x'$$

Limiting time and length by Planck-time and Planck-length and transforming translations in three-dimensional space in an avant-garde way into rotary processes in space-time, results in complementary views on time, length, speed, and space-time curvatures of the general theory of relativity GTR: Quantum mechanics gets its feasible "space-time view" on Heisenberg's uncertainty principle. A rotary set-up of space-time reveals interchangeable space-time dimensions and an overall super symmetry of space-time. It allows the discussion of the special theory of relativity and the general theory of relativity in combination with all quantum mechanical aspects, mass generation, and dark energy sources of an expanding universe.

II) Complementary views on space-time

For the following discussions, subjective simultaneity of events defines the perceived space and is defined by all events throughout space that fall into one single Planck-time of an observer's individual subjective time. Three-dimensional space of this observer is constructed by simultaneity across distances. Original x-coordinates of relativistic mechanics with a Cartesian system describe in the following discussion the distance between any two points in three-dimensional space that appear with simultaneity of events. x'-coordinates stay on a chosen x-line. Let us postulate that there cannot be any coincidence of two or more events within one Planck-time in the same observer's subjective space location within one single Planck length and combine this postulate with the proven concept of the constancy of speed of light, implying dimensional stability and process continuity within a moving inertial system. This complementary view causes remarkable changes.

Postulating that there cannot be a coincidence of simultaneous events in the same space location that is defined by a subjective single Planck length during one subjective Planck-time of an observer, any expansion or any shrinking of a homogeneous and isotropic space can only be explained by the corresponding change of simultaneity of events, if space is defined by all observable areas with a simultaneity of events. An event is defined by one single quantum leap of time or length. Due to the formulas of relativistic mechanics, any remaining observer in a starting location actually reads a different time along x' of any moving object. The term

$$\Delta t = \frac{v}{c^2} \frac{x_2' - x_1'}{\sqrt{1 - v^2/c^2}}$$

of the formulas of relativistic mechanics describes different time in different locations along x' of the moving object. Neglecting the non-linear relativistic effect, a time span is given by the reduced formula $\Delta t = \frac{v}{c}(x_2'/c - x_1'/c)$: An observer in a moving object will further on register simultaneity of events across this inertial system at any relative speed. The remaining observer in the starting system, however, will read sequential events instead, spread over the period Δt. This period increases linearly with the linear increase of relative speed. The picture changes, however, if subjective space is defined by the simultaneity of events that construct this space. In this case, $\Delta x'$ will be successively replaced by time intervals with a leaping function on Planck length and Planck time level, if length and time are limited at these values:

Δx' of the moving object is transformed into a time interval, and this interval is dilated just like any other observed time interval at relative speed.

In a reduced formula, the replacement quantity of Planck lengths by Planck time intervals depends only on the ratio v/c of relative speed and maximum possible speed of light. Reaching theoretically speed of light, simultaneous events along x' of the moving system are completely sequential events in the starting system. This non-simultaneity replacement of length by time can be visualized with the following rotation picture of figure 1, looking in there only on Planck length x-axis, on Planck time y-axis, on the broken line, and on the linearly increasing speed ratio on the y-axis from v/c = 0 up to v/c = 1. Considering no motion below the level of one Planck length and one Planck time because of being the minimum levels for any process, any speed increase shows always leaps, and the maximum acceleration a_{max} equals c/t_p. Planck length and Planck time cause the grey areas of figure 1 below Planck length and Planck time levels, allowing deviating positions of zero points of superimposed diagrams within the limits of one Planck length and one Planck time.

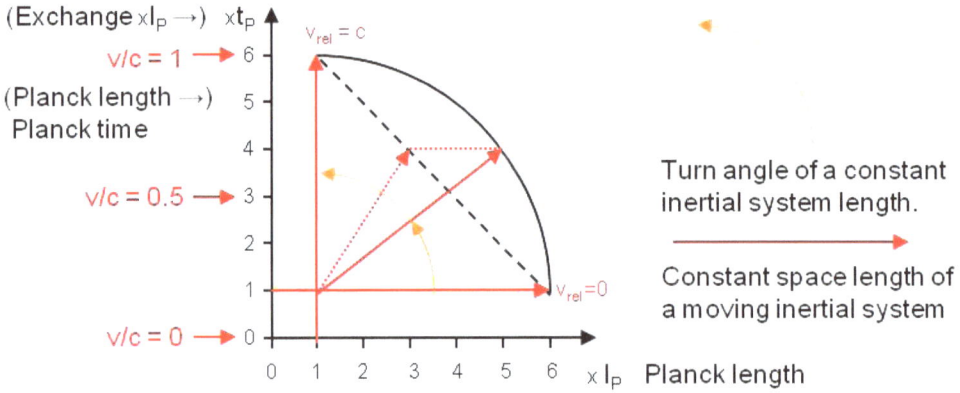

Figure 1: Rotation of a constant length into time by simultaneity in space

Any observer that remains in the starting environment would read for the doted length arrow of a moving object 3 Planck lengths by projection on the original length scale. This picture, however, cannot be correct, because it would imply a reduction of the moving system from originally 6 Planck lengths down to the length of the rotated doted red arrow between the zero point and the broken line calculable by $\sqrt{3^2 + 4^2} = 5$. This means that the step by step replacement of one Planck length by Planck time and treating them in the diagram equally by scaling each Planck time with a speed of light factor or scaling length by a speed of light divisor would lead to the subjective length reduction of the moving object for an observer in this moving inertial system and in the demonstrated case by exactly one Planck length, having the impact that light would disperse across this reduced length faster. This is in contradiction with discoveries of physics. 6 Planck lengths will stay constant in the moving system, being the basic prerequisite for the constancy of speed of light that is observable from any observation post. This proven fact can be easily integrated into figure 1, but keeping for the moment a linear y-axis unchanged, and according to a speed caused length-time replacement of a simultaneity area by the reduced formula $\Delta t = \frac{v}{c}(x_2'/c - x_1'/c)$, considering a superimposed factor $1/\sqrt{1 - v^2/c^2}$ upon the distance of $(x_2' - x_1')$ that keeps an original length of the x-axis constant during the rotation process in figure 1, because of keeping for example the 6 Planck lengths on the 6-unit-bow from zero speed up to the speed of light, not considering the leap functions and successive replacement of length by time for evaluations in the starting base. Figure 1 shows the replacement of length by time, but with a constant Δx': the increase of Δt is slowed down during the rotation of Δx' in space-time by the relativistic factor.

Keeping the dimensions of the moving inertial system and a starting system constant develops the circular bow in figure 1 to identify exchange leaps and to capture non-linear developments with stable linear x-axis and y-axis of a starting inertial system: In case that an original length of 6 l_P of the length-time exchange on the y-axis is re-calibrated linearly from 0 speed up to the speed of light c, figure 1 explains the relativistic length contraction $X\, l_P = \text{integer}\,(6\, l_P \sqrt{1 - v^2/c^2})$ for X = 1,2,3,4,5,6 according to Pythagoras formula $6^2 l_P^2 (\frac{v^2}{c^2}) + X^2 l_P^2 = 6^2 l_P^2$ and consistent with STR but now with leaps that are necessary to separate simultaneous events in the subjective space of simultaneity. This length contraction is derived by projection on the x-axis.

Any observer of the inertial starting environment cannot shift a subjective space-time position within this diagram but stays in a subjective central zero position for the start of consecutive events and measurement. This leads to an uncertainty in evaluations if cause and effect are not distinguishable, for example if the observer has left the starting position with relative speed, or if the starting position moves in relation to the observer. Figure 1 shows that any perception of simultaneity of events is subjective and that all space distances are generated by separation, if two or more simultaneous events cannot coincide in a space frame of a subjective Planck length within the interval of one subjective Planck time. Space may be thus interpreted as individual three-dimensional separation result for simultaneous events and any kind of space expansion would require a process that increases areas of simultaneous events at the expense of serial events, considering dilation and contraction effects, and other equivalent processes.

The important finding is the fact that simultaneity of events or sequential events could be both generated by an exchange of time and length and that this process can be mathematically and physically described by a rotary process in space-time, defining space as subjective area of simultaneity.

Figure 1 keeps the dimensions of the inertial starting system stable by the perpendicular and linear construction of the length-time exchange frame. However, the y-axis of a moving inertial system can be rotated together with its x-axis to show the correct developments and directions of relative time dilations of STR. This is demonstrated in the rotation picture 2. Note that the t_{R0D}-axis describes a continuous reference time of any inertial starting system. The measurement of relative time developments starts with setting t=0 on a clock in the moving object and on another clock that remains in the starting position.

Figure 2 shows the geometrical derivation of the relativistic time dilation: t_R describes the time of an object, moving with the speed $v_{rel} = v$ relatively to the initial system with time t_{R0D}.

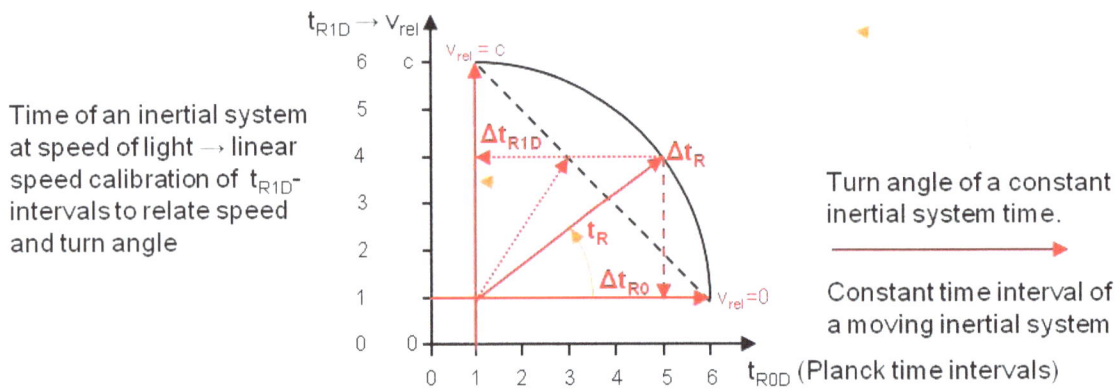

Figure 2: Relative time dilation in a moving inertial system

Anticlockwise t_R-rotation causes a relative t_R-dilation. The moving observer does not notice the dilation within the moving inertial system, but a faster pace of time in the original starting environment with its relatively contracted length frame. Figure 2 shows an example with the ratio 5/6 between the pace of t_R and the pace of t_{R0D}. The minimum of any subjective time is one Planck time $t_P = 5.391 \cdot 10^{-44}$ s, causing leaps in this space-time-speed diagram. t_{R1D} is a theoretical support axis, showing a relative standstill of time and an infinite dilation in relation to t_{R0D}. The linear speed calibration corresponds to the subjective possibility to linearly increase relative speed against a starting system. Not considering quantum leaps, this calibration is defined by

$$\frac{\Delta t_{R1D}}{\Delta t_R} = \frac{v}{c} \quad \text{or} \quad \Delta t_{R1D} = \frac{v}{c}\Delta t_R$$

The interval on t_{R1D} develops with the linear increasing speed relation factor v/c and Δt_{R1D} is defined by $\Delta t_{R1D} = Xt_{P1D}$, with $X = 1,2,3\ldots$ and Planck-time t_{P1D} of t_{R1D}. Δt_{R0D} is defined by Xt_{P0D} with $X = 1,2,3$ and Δt_R is defined by $\Delta t_R = Xt_{PR}$ with $X = 1,2,3$, introducing a relative Planck time t_{P0D} for t_{R0D} and a relative Planck time t_{PR} for t_R. Δ is henceforth indicating leaping functions.

Superposition of speed in a moving system upon the speed of the moving system itself leads in such a diagram to the non-linear speed addition of the STR with absolute speed of light barrier. Using the equation of Pythagoras for

$$\Delta t_{R0D}^2 = \Delta t_R^2 - \Delta t_{R1D}^2$$

results in

$$\Delta t_{R0D}^2/\Delta t_R^2 = 1 - v^2/c^2$$

The inverse equation describes an energetic prolongation of the time t_R to express for example the slower energy consumption by the relativistic factor in comparison with the original time frame if the calibration of time is not adapted according to the dilation:

$$\Delta t_R^2/\Delta t_{R0D}^2 = 1/(1 - v^2/c^2)$$

Considering the necessary re-calibration because of time dilation with the stretching of Δt_R results consequently in the inverse ratio for an observer on t_{R0D}:

$$\Delta t_{R0D}^2/\Delta t_R^2 = 1/(1 - v^2/c^2)$$

An observed time t_R is dilated in comparison to t_{R0D}, i.e. t_R runs in fact slower than t_{R0D}, though the moving object with time t_R does neither leap into the past nor into the future of t_{R0D} but stays continuously in the present, which has been proved in many experiments in circular accelerators.

It is possible to combine figure 1 and 2. The turn angles for an object have the same value. A straight movement in three-dimensional space has been transformed into rotary processes in space-time, if monitored from a central position of each event within a homogeneous, isotropic space. Although t_{R1D} is maximum dilated in relation to t_{R0D}, the rectangular construction of these two time axes seems to define the total segment of $t \geq 0$, in case that any negative time development into the past of events is forbidden, i.e. any development back beyond the grey areas. Relative shifting of the zero point of events within the Planck frames can generate immense effects, because of the involved relative time dilation or time contraction. Static pressure and dynamic acceleration of processes against t_{R0D} are capable to generate simultaneity of events and three-dimensional space. Figure 1 shows that the length of the moving object would fully coincide with the time axis t_{R0D} in case of reaching the speed of light. Defining additionally space by observable simultaneity of events transforms this length of the object into a t_{R0D} time interval that is dilated in correspondence to the relativistic length contraction. If the moving observer at the speed of light reads simultaneity across the vehicle, but the remaining observer in the starting environment reads sequential events within this moving vehicle, the picture suggests that we may read simultaneity of events throughout space because of being ourselves on another track in space-time at the speed of light. This would mean maximum dilation of our time t_{R0D} against another perpendicular time line t_{R3D} and passive speed of light on a fourth axis t_{R2D} that was originally a length but contracted into a Planck length because of the speed of light, and simultaneously rotated into t_{R3D}. If this assumption is valid for baryonic masses and their environment, this rotation process will generate space distances with simultaneity of events, and t_{R3D} changes into distances with the minimum of a Planck length. t_{R0D} is maximum dilated in relation to t_{R3D} to ensure this simultaneity. This way, all t_{R2D}- and t_{R3D}-processes can be completely described within the grey areas of simultaneity in figure 1 and 2, without any backwards running time in the completed picture with four axes $t_{R0D}, t_{R1D}, t_{R2D}, t_{R3D}$, with forward running t_{R0D}. t_{R2D} opposes t_{R0D} and t_{R1D} opposes t_{R3D}. Four axes that are rotated by 90 degrees are just the result of chaining consecutive and simultaneous events with individual relative zero points within the grey areas against other events, maintaining always $t_{R0D} > 0$.

A fourth axis t_{R2D} is simultaneously the 90 degrees turned speed calibrated time axis for all superimposed relative motions at active speed of light with t_{R1D} time, like for photons. Gravitation, inertia, and also three-dimensional space are in this model only possible because of the grey areas with the tension function of t_{R2D} and the maximum acceleration of t_{R3D}-processes against t_{R0D} that generates Planck lengths and space with simultaneity of events. Axis t_{R2D} may be calibrated for the subjective observation from t_{R0D} with a/a_{max}, introducing acceleration a and also maximum acceleration $a_{max} = c/t_P$ that describes acceleration from zero to speed of light in one single Planck-time interval. t_{R0D}, t_{R3D}, and t_{R1D} are capable to describe the special theory of relativity and t_{R2D} allows the realization of space-time curvatures of the general theory of relativity. Defining t_{R2D} additionally as an equal rotated quantity to t_{R0D}, t_{R1D}, t_{R3D} leads to quantity replacement and dilations also between t_{R1D}, t_{R2D}, and t_{R3D}.

III) Discussion of space-time leaps and conclusions

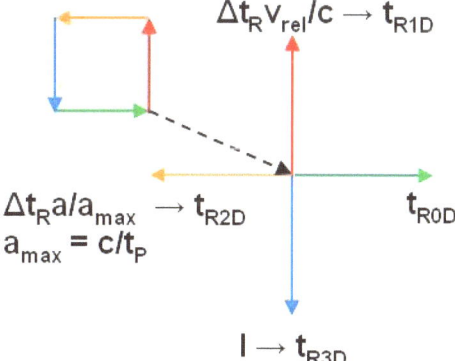

Figure 3: Space-time-speed-acceleration leap frame

The space-time-speed-acceleration leap frame as shown in figure 3 should have following features in order to entirely conform to relativistic mechanics and quantum mechanics:

The frame that is defined by subjective Planck time and Planck length has a central location for any event in space-time, in case of homogeneous and isotropic development of space-time. It has a defined dimension for events, and differs in location, relative size and orientation in comparison with other frames, generating simultaneous and serial events by respective functional alignment of many frames, and depending on the nature and location of an observer. Each of the four components of a space-time-speed-acceleration leap frame can change subjective functions for an observer only together with the other three components of its frame. It is the physical basis for the capability to build up three-dimensional space by spatial separation of all simultaneous events, and four-dimensional space-time with serial events along one, subjective time line. The rotation processes with the exchange of space-time-speed-acceleration quantities limit relative speed at the speed of light barrier, avoiding any relative standstill of time with respect to four rotated frame perspectives.

A leap-frame gets conclusive evidence, considering time, length, speed, acceleration to be combined and rotated manifestations of identical types of energy components, developing in clusters either propulsion or tension energy that can be cut down to an energy quantum of action[9] but not any further. Evaluation of gravitational features of t_{R2D}-acceleration components without capturing additional exchangeable features of t_{R0D}-, t_{R1D}-, and t_{R3D}-components misses 75% of total energy contributions that can increase the space area of simultaneity. Any long distance turns of accumulated frames cause impacts, depending on turn angles across clusters of frames.

Any relative rotation of elementary frames by 180° is supposed to generate stable negative acceleration densities with the appearance as various forms of matter in space-time because of the description in the following picture 4, in case of keeping their original parameters and quantities despite of such a rotation.

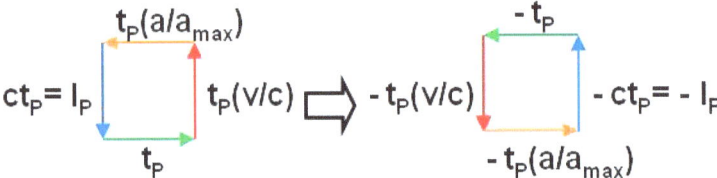

Figure 4: 180°- turn of space-time-speed-acceleration components

Taking the four units of frame components as observed from t_{R0D}, figure 4 shows the possible reasons for the generation of matter: a 180° rotated combination will appear with negative gravitational acceleration towards this rotated element. Such elements seem to be generated in pair as (180°)-matter and as (-180°)-antimatter by appropriate physical cause and effect scenarios. A confrontation of ±180°-rotated units with their original space-time environment causes a contraction of surrounding space by $-l_P$, and the time dilation by $-t_P$. Spatial aggregations of such units accumulate negative space-time quantities, increasing strength of gravitation and time dilation in the surrounding environment.

The rotation concept implies that the observer with subjective time t_{R0D} moves unnoticeably at the speed of light on the length l_{R2D}, causing the simultaneity of events on t_{R3D} and three-dimensional l_{R3D}-space instead of sequential events of t_{R3D}. The idea requires two different kinds of speed of light, an active speed of light, and also a passive speed of light. An initiated process with an active speed of light leads to evaluations of static features relatively to the observation base and complete transformation of sequential events into simultaneous events. Passive speed of light of the observer is system immanent and can be identified by the simultaneity of events in subjective space and by the particle and flow features at the speed of light. A dualism of wave and particles seems to be based on active and passive speed of light processes. This system immanent passive speed of light on l_{R2D} does not conflict with the fact that baryonic masses cannot reach speed of light on l_{R3D}: The fundamental problem of a 180°-rotated unit of not being able to reach the speed of light can be derived from figure 4: The acceleration component has replaced the Planck time component and a relative speed is now dilating this acceleration component just like any time component. This negative acceleration component reflects the mass of the rotated unit. Therefore, this mass is expected to increase its mass with the relativistic factor. Both speed of light processes, active and passive, are combined, because of linked space-time-speed-acceleration frames.

All events have to be unambiguously assigned by cause and effect in order to correctly calculate relative developments and changes by accumulated frames for any observer with a fixed location in space-time. Bouncing back of electrical charged particles causes a probability of locations of particles, because of the negative parameters at Planck-level and interchangeable quantities of length, time, speed, and acceleration. Probabilities of particle locations can be captured by the wave equations of quantum mechanics. A static electrical charge and dynamic magnetic momentum can be explained by acquired angular leap frame momentum during the particle generation by rotation and precipitation, and on base of active and passive speed of light aspects. Rotation of two ±90° frames by ±180° is supposed to produce electrically and magnetically neutral pairs of elementary particles, because of electrically and magnetically neutral angular leap frame momentum.

IV) Definitions for the construction of a relativistic time suite

- Vacuum of our universe consists of infinitesimal energy components that produce an observer's subjective individual four-dimensional space-time.

- Passive motion on one chained energy fabric dimension at the speed of light results in a manifestation of space with simultaneity of events.

- Active speed of light processes generates sequential events on relatively resting locations in space-time that can be defined areas of simultaneity.

- Any type of an infinitesimal energetic process should be describable by relative rotations in space-time-speed-acceleration event diagrams, or by relative dilation or contraction of all four axes of such diagrams.

- Combined symmetrical energy processes show an asymmetry, if they are monitored from an asymmetrical position in space-time, for example with a passive movement at the speed of light within the space-time energy fabric.

- Space-time flow and tension scenarios can be simplified by trigonometric projections of proportions between an object's time development t_R and a basic system time development t_{0D}, using a reference time interval t_{0F} and unchanged calibrations of time and length, despite time dilations and length contractions. This includes the gravitational aspects.

- The trigonometric approach simplifies considerably the energy tensors of GTR, because it describes one subjective single event in space-time with the continuous present of related objects.

- A time dilation of the special theory of relativity SRT can be geometrically explained with a graph, showing two perpendicular time axes t_{0D} and t_{1D}, and a turning object's time line t_R between those two.

- Any straight timeline t_R shows a nonlinear increasing angle against the straight timeline t_{0D} depending on the relative speed of the object with time t_R in comparison to its inertial starting system with time t_{0D}.

- A monitored t_F-time frame with an increasing dilation develops as an arc-function between t_{0D} and t_{1D} because of the constancy of speed of light and structural dimensions of the moving system. Therefore, it is possible to capture rotations by trigonometric functions and reference time interval t_F, if the time dilation is kept in mind, or substituted by the picture of relativistic energy storage.

- The t_R-dilation works in mutual directions between inertial systems, but depending on the relative acceleration history of these systems on base of cause and effect. The acceleration component of a space-time-speed-acceleration frame clarifies this direction of dilation, unambiguously. Not knowing the origin of relative motion and its acceleration history leads to the uncertainty principle of impulse power versus exact location around subjective fixed locations, because of not being in a position to identify whether simultaneous events will change into serial events, or vice versa. The very same identification problem arises in fixed space-time positions without a possibility to capture the percussions on this location, being the effects and not the causes. Only the time-length-speed-acceleration leap frame can achieve the rotary effect in both directions of two observed objects, but keeping all subjectively perceived dimensions and the pace of time on each of both object constant, including a constancy of speed of light.

- An increase of the relative speed rotates the time line t_R of any object away from t_{0D}-heading until it reaches a right angle at the relative speed of light c. The perpendicular set-up from zero is merely a mathematical support construction with a relative time development t = 0. It is important to repeat that those grey areas of figure 1 and 2 have to be existent for every consecutive event in case of the generation of continuous time. In this case, the perpendicular construction of all four axes is acceptable.

- Trigonometric time-projections will be always related to subjective and individual t_{0D}-lines, i.e. to the defined time of an original inertial system.

- Starting any measurement and comparison of time will get a zero point at the start of each measurement. According to relativistic mechanics this is possible if the time intervals in the moving object are measured in fixed space coordinates of this moving system.

- $l_P = c \cdot t_P$ is valid with Planck's length l_P and with Planck's time interval t_P, $l = c \cdot t$ is valid, with prolongation of l_P to $l = k \cdot l_P$, and prolongation of $t = k \cdot t_P$. $k = 1,2,3,4, \ldots$ are integers, c the speed of light in a vacuum.

- The segmentation by perpendicular time axes will be called "Relativistic Time Suite", abbreviated RTS, showing rotary super symmetry of space-time-speed-acceleration processes and the principle of cause and effect.

- Time axes are summarized in some of the formulas as $t_{XD}(X = 0,1,2,3)$. They portray the perceptions of time, length, speed, and acceleration.

- These straight timelines can be fully turned into each other in the order: $t_{0D} \leftrightarrow t_{1D} \leftrightarrow t_{2D} \leftrightarrow t_{3D} \leftrightarrow t_{0D}$. They are kept straight as well as the space lengths. Space-time curvatures of GTR, causing for example deviation of light paths by gravitation are captured by rotations in the diagram. These rotations conform to GTR and in sufficiently small areas of space-time or in case of a mass free universe to STR. Stringing leap processes of time and length components together results in the curvatures of GTR.

- Development of the system time t_S of any observer shows $\Delta t_S \geq t_P$ with Planck time $t_P = 5.391 \cdot 10^{-44}$ s that limits the grey areas of figure 1 and figure 2.

- All four "S-TQ" ("speed time quartet") time directions develop relatively to each other with subjective individual $t \geq t_P$: The t_P-leap-frame appears for any observer on a subjective timeline as tidal energy quantum foam fabric because of having obtained functional differences. The geometrical figure of such a multidimensional t_P^4-sphere is named c_{T0}.

- In order to visualize different space-time positions and developments in relation to $t_{0D}, t_{1D}, t_{2D}, t_{3D}$, it is possible to project the four time axes onto a single plane, rotating the time axes by 90°, as shown in figure 5.

- This simplification causes opposing time directions, actually resulting in a three-dimensionally spreading space. Only three-dimensional t_{3D}-escape routes avoid any coincidence of simultaneous energetic events in a four-dimensional time-length-speed-acceleration scenario with center location of each single event in case of homogeneous time and homogeneous and isotropic space.

- It is possible to differentiate space-time segments by subjective different appearances as time and space or static features at active speed of light and dynamic features at passive speed of light. Figure 5 shows these classifications.

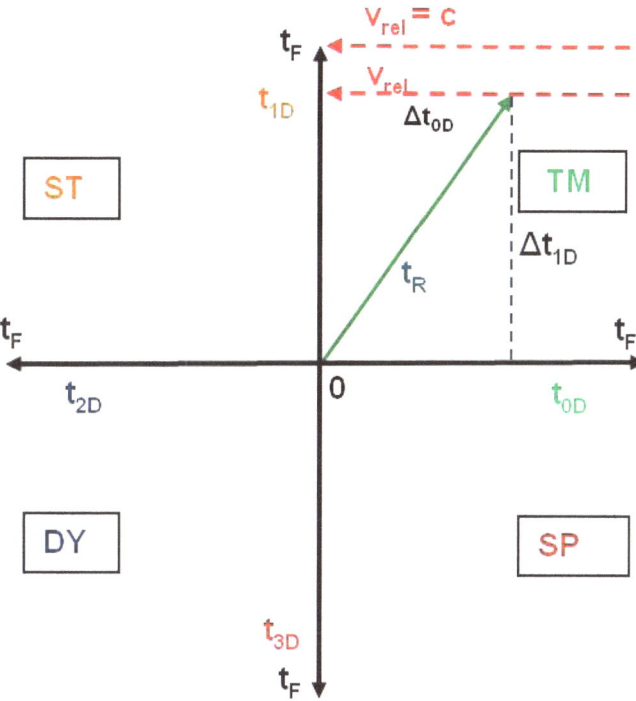

Figure 5: Relativistic dimension segments

The time sector between t_{1D} and t_{0D} gets the name **TM** for time and the sector between t_{3D} and t_{2D} is called **DY** for the dynamic, passive speed of light area. The static active speed of light area is called **ST** and borders on t_{2D} and on t_{1D}, whereas three-dimensional space **SP** borders on t_{0D} and on t_{3D}.

- A subjective evaluation on any t_R-line is always showing a complete and right-angled, but in relation to t_{0D} dilated S-TQ-set. This is a fundamental prerequisite for the constancy of speed of light and for a linear t_{1D}-speed scale from each t_R-line. This results in non-linear superposition of relative speed in case of adding speed measurements from the starting position with speed measurements within relatively moving inertial systems.

- Crossing the active speed of light barriers should add consecutively one dimension of simultaneity. Therefore, t_{3D} adds up as a three-dimensional distance vector in space. Crossing passive speed of light barriers seems to affect three dimensions.

- Having an observation point with sequential events on t_{0D} and passive speed of light on l_{2D}, we interpret the symmetrical picture differently: the four time axes do not any more appear as time lines. t_{0D} and t_R appear as one-dimensional time thrust in any definable point throughout space, t_{3D} shows up as a space length, t_{1D} is experienced as a speed barrier, and t_{2D} appears with acceleration features. The gravitational constant G, experimentally derived with the value and unit:

$$G = 6.67428 \; (+/- \; 0.00067) \cdot 10^{-11} \mathrm{N} \cdot \mathrm{m}^2/\mathrm{kg}^2, \text{ i.e. } \frac{\mathrm{m}^3}{\mathrm{kg} \cdot \mathrm{s}^2} \qquad \text{(Codata 2006)}$$

calibrates masses with unit kg within this arrangement, and, if applied to the pure relative acceleration value of t_{2D} only, with kg/m^2.

- A gravitational time dilation can be explained by reduction of the t_F-value within an S-TQ-diagram, as counter reaction to an increasing aggregation of masses. This contraction corresponds to the space-time-curvatures of general relativity, but simplifies the processes. t_{0D} and t_{3D} show equal dilation directly with the result of a relative t_F-contraction and according length contraction. t_{1D} builds up tension, but from a perspective within the t_{0D}-flow, with a constant relativistic ratio $\Delta t_{3D}/\Delta t_{0D} = \Delta t_{3D}'/\Delta t_{0D}' = 1$. This indicates that the speed characteristics in moving systems stay within this system just as they were without relative motion. Therefore, t_{1D} may be calibrated from any subjective t_{0D} timeline with linear development from 0 up to the speed of light for any chosen reference length of t_F and t_R. t_{2D}, however, shows its peculiar inversion from the viewpoint of time channel t_{0D}: The non-relativistic t_{0D}-view of Δt_{2D} as ratio $\Delta t_{3D} \cdot c / \Delta t_{0D}^2$ with unit m/s^2 and space-time-ratio $\Delta t_{3D}/\Delta t_{0D} = 1$ causes the inverted perception $\Delta t_{2D} \sim 1/\Delta t_{0D}$: In case of t_{0D}-dilation, i.e. relative contraction of Δt_{0D}, we observe an increase of corresponding gravitational acceleration. Keeping additionally the subjectively observed Δt_{3D}-space frames constant results in: $\Delta t_{2D} \sim 1/\Delta t_{0D}^2$, showing Newton's non-relativistic picture of gravitation: $(\Delta t_{3D} \cdot \Delta t_{0D} =) \Delta t_{0D}^2 \sim 1/g$. Result of all these considerations: the non-relativistic observations within t_{0D} at an assumed basic speed of light can be transformed into S-TQ-processes, with its perpendicular timelines.

- Considering effects of time dilations, the velocities of different objects can be non-linearly added with superimposed S-TQ-scenarios: S-TQ-pictures of the moving objects have to be first rotated and then superimposed with respective dilated timelines. References to relatively longer timelines t_{0D} are deciding about dilation impact of several connected moving scenarios on each other, according to interconnected history of objects and events, causes and effects. A longest timeline t_{0D} makes only sense in case of a one-way unbalanced picture of t_{0D}-cause and t_R-effects.

- The differential geometry of any observable space-time scenario leads to different evaluations, if this scenario has been built up from t_{0D}, from t_{1D}, from t_{2D}, or from t_{3D}.

- Quantum leaps during segment transitions seem to cause deflections of energy components into new t_{XD}-directions with appearances and effects of rotating vortexes, caused by residual momentum and detectable in form of spins and electrical charges of elementary particles. Electrical charges show the static behavior of segment ST and magnetic features of spins could be classified as dynamic processes of segment DY.

- Segment ST is supposed to represent electrical characteristics, segment DY magnetic characteristics. Processes of ST and DY spread together with active and passive speed of light, perpendicularly to each other.

- Angular momenta of anterooms appear as right-angled energy stream superposition within the energy stream of a subjective observation space.

- Therefore, four different angular momenta of energy streams may initiate under certain circumstances the construction of elementary particles.

- Elementary knots of tied component detachments differ from each other because of the individual sequence of linking and rotation, forming matter and anti-matter.

- An assumed construction of matter with angular momentum of anterooms may be the basic reason for matter waves to develop according to partial differential equations of second order in location and of first order in time.

- The SP-segment develops a three-dimensional LWH-space with length, width, height because of the four t_{XD}-dimensions, if one t_{XD} manifests as the subjective system time t_{0D}, based on the relativistic transformations of sequential events of t_{0D} and simultaneous events of three-dimensional t_{3D}-escape routes. The reason seems to be the passive speed of light on l_{2D} generating simultaneity of t_{3D}-events with three-dimensional escape routes that form our space with the center location of any observer.

- Figure 6 summarizes static momentum and dynamic angular momenta in all four space-time-speed-acceleration segments that appear for example as electric charges and magnetic quantum, depending whether they have been acquired from a relative active speed of light segment as a single static momentum or from the relative passive speed of light segments.

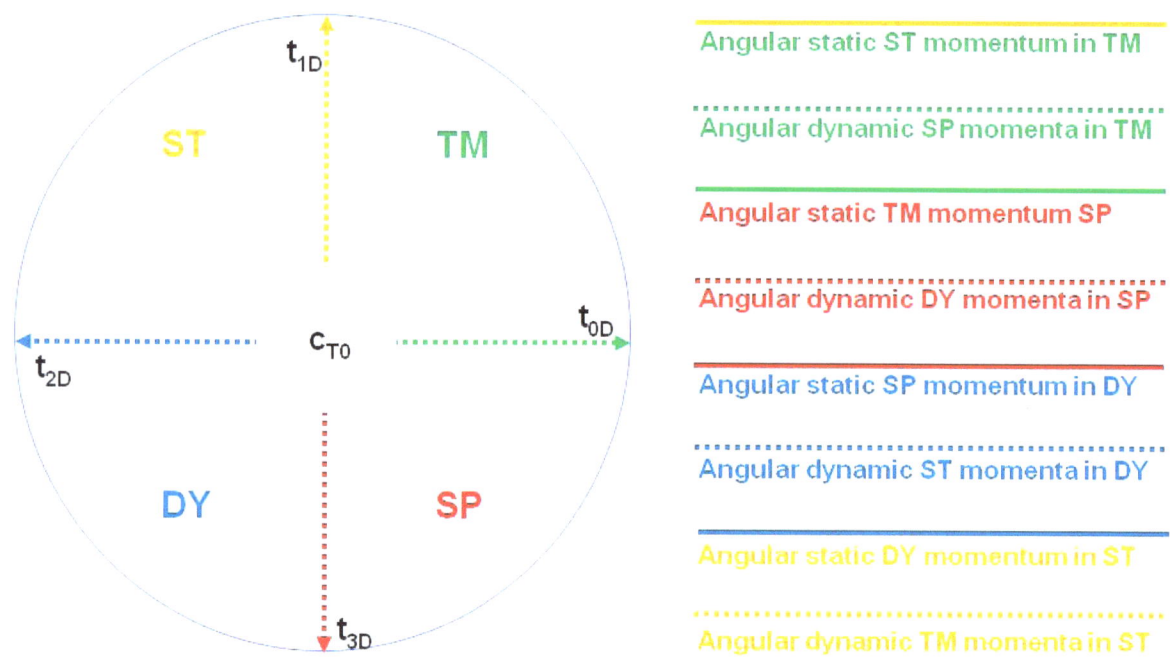

Figure 6: Static & dynamic angular momenta by active and passive speed of light

- Only relativity and quantum mechanical break up allow that four basically sequential time elements develop right-angled and with rotary symmetry.

- The t_{0D}-development shows not a spatial dimension, but is developing in observation points throughout space as time parameter in any definable spot in a subjectively experienced space. This is the reason for index 0D.

- Figure 7 shows expected areas of particle generation. Positive particles in free form represent antimatter in SP and lead to in-pair exterminations. Regular atomic SP nuclei are positive and supposed to be generated by distorted space-time elements but stable core constructions. Antimatter with negative charged nuclei has been already artificially produced.

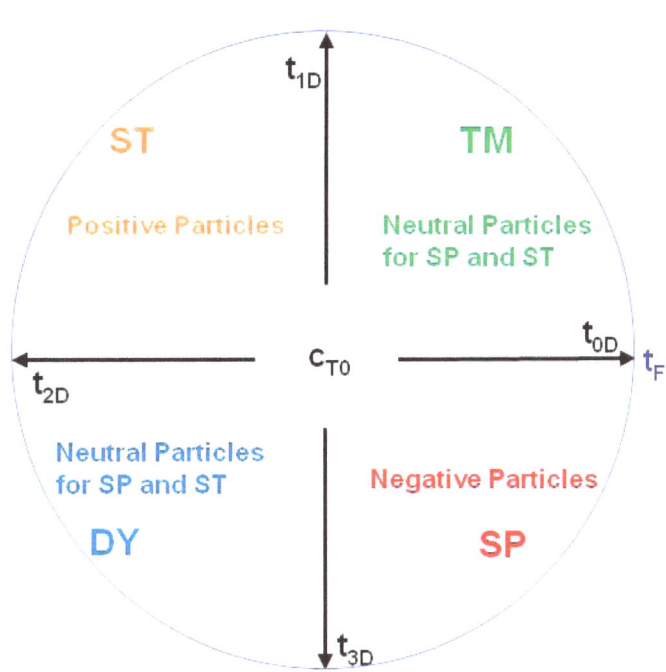

Figure 7: Expected areas of free particle set generation by cause and effect

- Electromagnetism seems to be based on ST with static of electricity and DY with the passive speed of light features of magnetism.

- Collective alignments of angular momenta and of their spatial movements cause coupled, rectified electromagnetic fields. A generation of magnetic fields and electric fields are technical alignment applications, with spatial dynamics of magnetism and basic static behavior of electricity.

- Relative speed means relative slowing against t_{0D} relative to observation posts by $-\Delta t_{0D}/t_{0D}$, according to figure 8 and an increase of the t_{1D}-component by $+\Delta t_{1D}/t_{1D}$. This increase of t_{1D}-shares requires spatial movement within SP-space.

- The time processes in ST came to a standstill in relation to processes in TM, processes in TM in relation to processes in SP, processes in SP in relation to processes in DY and processes in DY in relation to processes in ST. Therefore, electromagnetic waves can be transmitted at the speed of light without any losses of superimposed information across long distances in TM-time and in SP-space. This is the base for all static and dynamic coupled active speed of light processes and passive speed of light processes with multifarious appearances in nature.

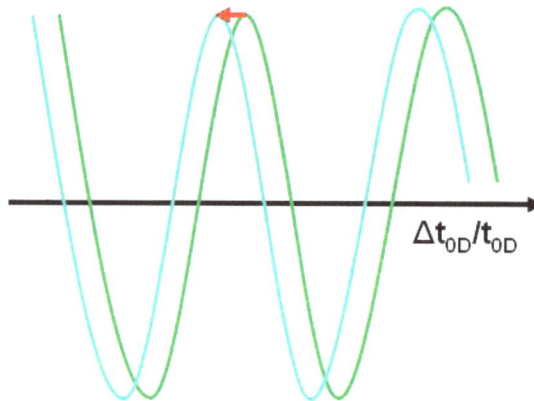

Figure 8: Triggering a relative wave as response to an excited oscillation

Observing from t_{0D}, the speed v_{rel}, generates a relative wave against any base t_{0D}-oscillation because the time t_{0D} cannot run backwards and the object moves with v_{rel} slower through subjectively experienced, original t_{0D} time.

- Energetic, geometrical compensation points and areas are scaffolding for elementary particles and all space developments, forming a geometrical equilibrium quantum foam figure named "c_{T0}" for all definable space-time-spots.

- Trigonometric discussions of S-TQ-diagrams reflect the general theory of relativity for simple scenarios. Using absolutely equal scales for time, i.e. not considering any dilation, the trigonometric discussion of the ST-Q-diagram is possible because the length of a time interval indicates the proportional relative energy content: A battery pack on any timeline will last exactly the same subjectively experienced time period, but the actual consumption of energy on rotated lines is slowed down in comparison with the consumption on the initial starting line because of time dilation. Example: Δt_R in figure 1 increases relative energy in relation to Δt_{R0D}, proportionally to time dilation. One result is the relative gain of mass by relative speed. Any excitation by electromagnetism or by a motion from individual starting line t_{0D} causes changes in the embedding space-time environment. This distortion could be captured by trigonometry for these simple space-time scenarios.

- t_R-time of moving matter assemblies can be mathematically related to a basic t_{XD}-constellation and also to each other with the support of four scaffolding timelines t_{XD}, and by means of four cosine functions.

- The trigonometric formula for two ideally perpendicular time directions t_{XD} and $t_{(X-1)D}$ is:

$$t_{XD} = t_F \cdot \cos(\pi/2 \cdot \vec{v}_{Xrel} \cdot t_{(X-1)D}/t_F \times \vec{1}/c_{XR} + \varphi_0)$$

X = 1,2,3,4, and with $t_{4D} = t_{0D}$, scaling arguments with $\pi/2$.

- This formula is valid for all four t_{XD}-sectors, just as it would be perceived within the t_{0D}-t_{1D}- sector. If the two introduced vectors $\vec{v}_{Xrel}$ and $\vec{1}/c_{XR}$ are perpendicular to each other, results:

$$t_{XD} = t_F \cdot \cos(\pi/2 \cdot v_{Xrel} \cdot t_{(X-1)D}/t_F \cdot 1/c_{XR} + \varphi_0),$$

with the angular velocity:

$$\omega = \pi/2 \cdot v_{Xrel}/t_F \cdot 1/c_{XR}.$$

- Observed from the position of any time vector t_{XS} of a matter assembly, it is valid $v_{Xrel} = c_0$ = constant, setting $\varphi_0 = 0$ and with axiom $t_{XD} > 0$. $\pi/2$ of the argument takes time and velocity onto the circle line $2\pi \cdot v_{Xrel}$, and reduces the relativistic observation phase to the $\pi/2$-sector, in which t_{XD} has started from a defined zero point and reached a defined reference time t_F on $t_{(X-1D)}$. These considerations indicate a possibility to treat time lines as a new form of energy with reactive constructions of perpendicular headings, building up flow and tension scenarios.

- This formula has the advantage that t_X-time-dilations can be discussed by relative speed and gravitational dilatation can be analyzed by changes of t_F-reference intervals, i.e. with comparable results, but distinguishable intervention parameters.

- The vector $\vec{c}_{XR}^{-1}$ is perpendicular to the vector $\vec{v}_{Xrel}$, maintaining the static ante-room direction. Static "Ante-room" is for TM: ST, for ST: DY, for DY: SP, and for SP: TM. Dynamic "Follow-room" is for TM:SP, for SP:DY, for DY:ST and for ST:TM.

- It is more feasible to take the "sliding base" $\vec{c}_{XR}^{-1}$ of the static anterooms instead of dynamic follow-rooms because of their permanent indifference. Moreover, initiation of active speed of light processes seems to be only possible by harsh deceleration processes that cause the excitation of the static anteroom via the dynamic follow-room direction.

- Time equations vary in their phases: in relation to the TM-equation, the equation of SP is out of phase by +90° or -270°, of DY by ±180°, of ST by +270° or -90°.

- A static slipstream-curve which portrays motion of an assembly of matter with the time vector t_R, starting on t_{1D} and heading towards t_{0D}, can be described by the equation: $t_{1D} = t_F \cdot \cos(\pi/2 \cdot t_{0D}/t_F)$. This trigonometric formula is describing the movement of the crossing point of π/2-arc and radial S-TQ time vector t_R. The time t_R of a sufficient large assembly of matter gets time indicator t_S, in order to define and to design independent inertial systems and to compare those with each other. The arithmetic middle of earth time is described by t_E.

- The extension of space-time areas in the basic formula can be limited by t_F. Therefore, single particles or particle associations that are generated by partial rotated areas can be separated in space-time and moved away from matter assemblies with relative speed.

- Relative speed v in relation to v_{Xrel} may be described with : $v_{Xrel} = c - v$. Note the deduction of the relative object's speed from the speed of light level and that this relatively moving object is further on always carried in the present of the starting time line: the time dilation and a capture in the present of the original time has to be kept in mind, using trigonometry.

- The basic speed level within any of the four t_{XD}-streams is speed of light. Subjective evaluation with any t_{XD}-time will show an asymmetrical picture of interacting flow and tension energies.

- Relative velocity changes the phase angle φ_0 against original positions with $\Delta\varphi_v = -\pi/2 \cdot v \cdot t_{XD}/t_F \cdot 1/c_{XR}$, initiating a relative wave with slower frequency: time runs in fact relatively slower.

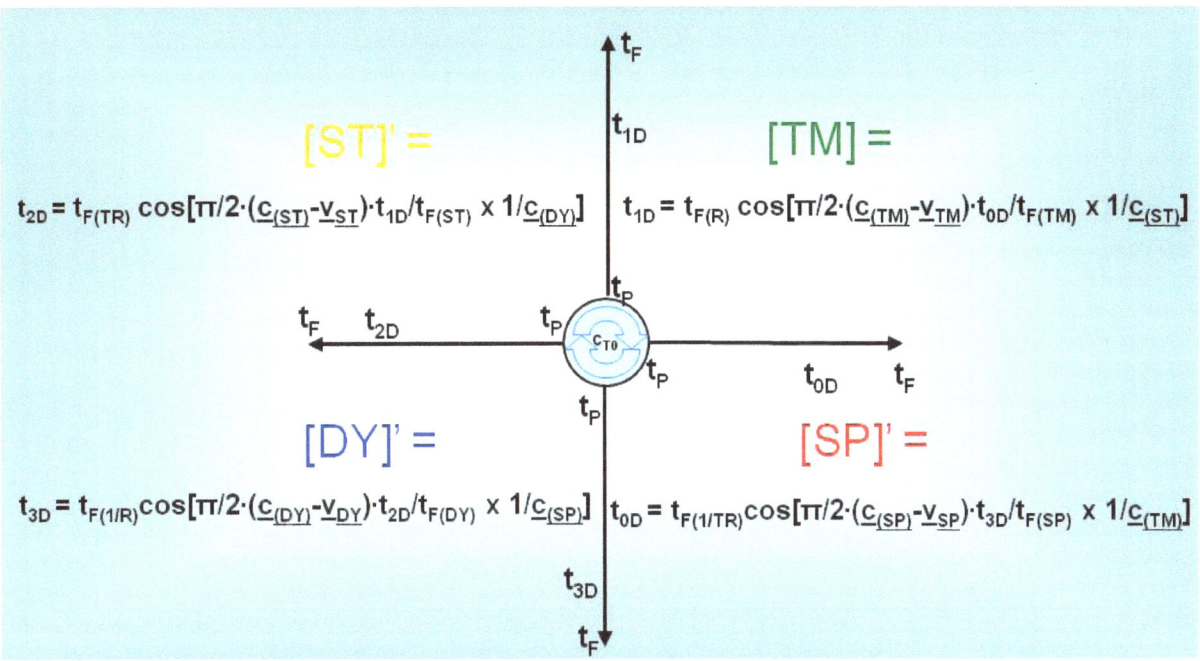

Figure 9: Basic t_{XD}-equations of all four sectors

The equations have not yet been related to each other. Linking of the functions has to take into account the different function phases against each other. All four trigonometric functions are related to the reference time interval t_F.

- Coupling of the four t_{XD}-equations causes processes of quantum phase leaping because of the system's component constraint $t > 0$.

- Linking of time functions with relativistic four time directions leads to the following equation, with the integer, quantum mechanical leap and twist-off-numbers z:

$$E_Z = \{\vec{t}_{F(1D)} \cos[\pi/2\, (\vec{c}_{(TM)} - \vec{v}_{(TM)}) t_0/t_F \times \overrightarrow{z_{(ST)}/c_{(ST)}}]\}$$
$$\times \{\vec{t}_{F(0D)} \cos[\pi/2(\vec{c}_{(SP)} - \vec{v}_{(SP)}) t_3/t_F \times \overrightarrow{z_{(TM)}/c_{(TM)}}]\}$$
$$\times \{\vec{t}_{F(3D)} \cos[\pi/2\, (\vec{c}_{(DY)} - \vec{v}_{(DY)}) t_2/t_F \times \overrightarrow{z_{(SP)}/c_{(SP)}}]\}$$
$$\times \{\vec{t}_{F(2D)} \cos[\pi/2(\vec{c}_{(ST)} - \vec{v}_{(ST)}) t_1/t_F \times \overrightarrow{z_{(DY)}/c_{(DY)}}]\}$$

- Changing the sequence of cross product terms decides on the subjective kind of three-dimensional space and one-dimensional time.

- Changing t_3 into t_2 and t_1 into t_0 converts the cosine functions into sine functions, and relative time dilatations accelerate the sine terms with t/t_P

against t/t_F, turning flow characteristics into tension characteristics from the perspective of t_0 and time area TM.

$$E_{Zt0,2} = \{\vec{t}_{F(1D)} \cos[\pi/2\,(\vec{c}_{(TM)} - \vec{v}_{(TM)})t_0/t_F \times \overrightarrow{z_{(ST)}/c_{(ST)}}]\}$$
$$\times \{\vec{t}_{F(0D)} \sin[\pi/2(\vec{c}_{(SP)} - \vec{v}_{(SP)})t_2/t_P \times \overrightarrow{z_{(TM)}/c_{(TM)}}]\}$$
$$\times \{\vec{t}_{F(3D)} \cos[\pi/2\,(\vec{c}_{(DY)} - \vec{v}_{(DY)})t_2/t_F \times \overrightarrow{z_{(SP)}/c_{(SP)}}]\}$$
$$\times \{\vec{t}_{F(2D)} \sin[\pi/2(\vec{c}_{(SP)} - \vec{v}_{(SP)})t_0/t_P \times \overrightarrow{z_{(DY)}/c_{(DY)}}]\}$$

- Let us now project in a second step the t_{2D}-axis onto the t_{0D}-axis. The processes along t_{2D} distinguish themselves with inverted mathematically development in comparison to processes along the t_{0D}-axis. Therefore, inverted terms have to balance the E_Z-equation due to the earlier findings of $\Delta t_{2D} \sim 1/\Delta t_{0D}$. The resulting equation for the vector field E_{Zt0R} develops as:

$$E_{Zt0R} = \cos_{1D}[\pi/2\,(\vec{c}_{(TM)} - \vec{v}_{(TM)})t_0/t_F \times \overrightarrow{z_{(ST)}/c_{(ST)}}]$$
$$\times 1/\sin_{0D}[\pi/2(\vec{c}_{(SP)} - \vec{v}_{(SP)})t_0/t_P \times \overrightarrow{z_{(TM)}/c_{(TM)}}]$$
$$\times 1/\cos_{3D}[\pi/2\,(\vec{c}_{(DY)} - \vec{v}_{(DY)})t_0/t_F \times \overrightarrow{z_{(SP)}/c_{(SP)}}]$$
$$\times \sin_{2D}[\pi/2(\vec{c}_{(ST)} - \vec{v}_{(ST)})t_0/t_P \times \overrightarrow{z_{(DY)}/c_{(DY)}}]$$

Agreeing, that the trigonometric cross product terms define the relativistic directions of trigonometric functions, the amplitudes t_F may be cancelled. This is possible in case of equal dilations or contractions of all time axes. 0D, 1D, 2D and 3D describe the alignment of each trigonometric term, respectively.

0D, 1D, 2D and 3D describe the alignment of each trigonometric term, respectively.

- The E_Z-equation shows a distinctive feature:

All four trigonometric arguments develop between $(4z + 1) \cdot \pi/2$ and 0, and with $z = 0,1,2,...$, in order to maintain $t > 0$ between the four time directions. Crossing limits leads to the complete exchange of functions. Quantum mechanics is based on these leaps and functional exchanges.

The result is the leaping function at Planck-time-levels and a construction of relativistic time sectors, because of involved momentum and mutual tension energies.

- The acceleration features of coupled trigonometric functions leads to the observation that time functions are sort of twisted off in t_P-tensor-field-slices by the directly following function within the cross product equation.

- This consecutive t_P-twisting-off shows the effect that any t_P-leap causes a relativistic rotation of $4 \cdot \pi/2 = 2\pi$ into the original picture, i.e. 360° with $z'_x = z_x + 4$, however, at the expense of rotation energy. Contributions of the $\cos_{3D}$-function are especially remarkable: t_P-leaps of the consecutive second twisting off process with the original starting base at the $\cos_{1D}$-function appear as LWH-perception with l_P-lengths, building up space dimensions with increasing expansion velocities, if they can be observed across sufficiently long distances.

- These length leaps develop into radial directions in relation to any starting point of an individual event in space, with retrospective views from this starting point because of the speed of light delay of any energy and information transport.

- A time coupled speed $\vec{v}$ can be written into the equation:

$$E_{ZtoR} = \cos_{1D}\left[\pi/2\,(\vec{c}_{(TM)} - \vec{v})t_0/t_F \times \overrightarrow{z_{(ST)}/c_{(ST)}}\right]$$
$$\times\, 1/\sin_{0D}\left[\pi/2(\vec{c}_{(SP)} - \vec{v})t_0/t_P \times \overrightarrow{z_{(TM)}/c_{(TM)}}\right]$$
$$\times\, 1/\cos_{3D}\left[\pi/2\,(\vec{c}_{(DY)} - \vec{v})t_0/t_F \times \overrightarrow{z_{(SP)}/c_{(SP)}}\right]$$
$$\times\, \sin_{2D}\left[\pi/2(\vec{c}_{(ST)} - \vec{v})t_0/t_P \times \overrightarrow{z_{(DY)}/c_{(DY)}}\right]$$

- Indices of c-vectors indicate a respective relevant subspace. The indices of c^{-1}-vectors describe a static sliding base. An E_{ZtoR}-equation shows that the t_0/t_P-factors of the relative time developments and relative spatial movements accelerate processes up to passive and active speed of light, in the $\cos_{3D}$-LWH-space. LWH-space experiences its three-dimensional spreading against TM by this $\cos_{3D}$-term.

- Starting processes from defined time positions that are not coinciding with t_{0D} leads to the relative dilated system time t_S of such positions. It is the basic reference time t_S of a superimposed process.

$$\begin{aligned}E_{ZtoR} = &\cos_{1D}\left[\pi/2\,(\vec{c}_{(TM)} - \vec{v})t_S/t_F \times \overrightarrow{z_{(ST)}/c_{(ST)}}\right]\\ &\times 1/\sin_{0D}\left[\pi/2(\vec{c}_{(SP)} - \vec{v})t_S/t_P \times \overrightarrow{z_{(TM)}/c_{(TM)}}\right]\\ &\times 1/\cos_{3D}\left[\pi/2\,(\vec{c}_{(DY)} - \vec{v})t_S/t_F \times \overrightarrow{z_{(SP)}/c_{(SP)}}\right]\\ &\times \sin_{2D}\left[\pi/2(\vec{c}_{(ST)} - \vec{v})t_S/t_P \times \overrightarrow{z_{(DY)}/c_{(DY)}}\right]\end{aligned}$$

- A slowing down of time development within E_Z has been considered by the minus sign of the relative speed $\vec{v}$. The development of S-TQ can be compared with the inflation and simultaneous contraction of a big balloon with several opposing dimensions. Being located on the surface of such an inflating balloon, we experience expanding distances, and at the same time we increase depth in the subjective center, with retrospective views around.

- A classification of subdivisions is based on the cross products within the trigonometric terms and the sequence of term linking. The subdivision 1 generates c-propulsion with reducing influence in case of relative speed. Subdivision 2 describes the static sliding forces that are caused by 1/c-subdivision-alignment and possible c^{-4}-linking of angular perpendicular elementary momenta and precipitation into elementary particles.

- Advanced mathematic calculation methods are able to describe multiple, complex interacting E_Z-constellations. Superimposed, rotating fields can be visualized and calculated as results of spatial differential operations.

- Observing E_Z-energy balances with quantum mechanical leaps simplifies the E_Z-equation, using an E_Z-quotient:

$$E_{ZtoR} = \frac{\cos_{1D}\left[\pi/2(\vec{c}_{(TM)} - \vec{v})t_S/t_F \times \overrightarrow{z_{(ST)}/c_{(ST)}}\right] \cdot \sin_{2D}\left[\pi/2(\vec{c}_{(ST)} - \vec{v})t_S/t_P \times \overrightarrow{z_{(DY)}/c_{(DY)}}\right]}{\sin_{0D}\left[\pi/2(\vec{c}_{(SP)} - \vec{v})t_S/t_P \times \overrightarrow{z_{(TM)}/c_{(TM)}}\right] \cdot \cos_{3D}\left[\pi/2(\vec{c}_{(DY)} - \vec{v})t_S/t_F \times \overrightarrow{z_{(SP)}/c_{(SP)}}\right]}$$

- E_{ZtoR} is a product of tidal streams and tensions, including considerations of the mathematical term inversion in form of reciprocals and vortex.

Using the summarizing terms [TM], [ST], [SP], [DY] leads to transparent relations of a subjective E_{ZtoR}-constellation, provable by electromagnetic processes:

$$E_Z = \frac{[TM] \cdot [ST]}{[SP] \cdot [DY]}$$

- The exchange of static and dynamic functions between [TM], [ST], [DY], and [SP] ensure continuous t_P-patchwork. The appearances of positive and negative polarities are the result of in-pair production of particles.

- These polarities can be experienced by two opposing [SP]-[ST]-charging features, with point charge characteristics in [SP]-space, and as coupled [DY]-[TM]-processes of magnetism with passive speed of light features of magnetic momentum and magnetic flow in [DY] and magnetic tension in opposing [TM] that can be rotated in [SP]-space. As important result, attraction and repulsion scenarios in the SP-space and within TM-time-environment are possible for electricity and magnetism, and a screening off of these forces.

- The separation into positive charges and negative charges determines a rotation direction within E_Z-equations in case of relative speed, according to the rules of electrical engineering.

- The E_Z-equation shows the basic constellation of a super symmetry of space-time. Development and linking of E_Z-equations lead to complex solutions.

- If monitored out of a fixed observation point, tidal energy moves without relative speed anticlockwise in relation to the equation, i.e. in TM in the direction of SP, in SP in the direction of DY, in DY in the direction of ST, in ST in the direction of TM.

- This movement is forced because of t > 0, and it is slowed down because of four quotients c_{0X}/c_{0Y}. A relative velocity causes an overlaid rotation into the opposite direction. A starting point of relative speed is defined as zero point of reference for individual time and length comparisons that will show always system immanent increase of space, despite of storage in multifarious forms of acceleration densities and speed scenarios.

- The magnetic field constant μ_0 and the electric field constant ε_0 with the permeability μ_r or permittivity ε_r, respectively, describe ST – DY – SP – TM interactions and observations, without masses and with masses.

Fixing the magnetic field constant $\mu_0 = 4\pi \cdot 10^{-7}$ (V·s / m·A) by definition and rotary magnetic tension (Volt V indicates electric tension and Ampere A the generation of magnetic tension) results in an electric field constant:

$$\varepsilon_0 = \frac{1}{\mu_0 c_0^2} = 8.854187817 \cdot 10^{-12} \text{ (C / V·m)}$$

Equation and units confirm the inverse contributions of SP and DY, and transverse linking of ST and TM in the numerator, and SP and DY in the denominator, respectively. Comparing this equation with E_Z shows that in a single point of space-time, seconds and meters are ruled by the cosine-terms, and the Volts and Ampere by sine-terms, once the E_Z-equation has been completely related to t_{0D} or t_S. This means that it is possible to to vary electric and magnetic parameters from any single point of space-time during sufficiently long t_F-frames and sufficiently short time intervals linearly, because of sine-functions. On the contrary, seconds and meter will stay during long t_F-frames and short time intervals constant because of cosine-functions. Hence for non-relativistic environments meters define spatial distances that can be measured by the absolute constant frame of speed of light in the overall system, because of elementary leap frames.

- There are four perspectives onto any complete scenario: These energetic rotated constellations should be equivalent in function and perception:

$$E_{ZTM} = \frac{[TM] \cdot [ST]}{[SP] \cdot [DY]} \quad E_{ZSP} = \frac{[SP] \cdot [TM]}{[DY] \cdot [ST]} \quad E_{ZDY} = \frac{[DY] \cdot [SP]}{[ST] \cdot [TM]} \quad E_{ZST} = \frac{[ST] \cdot [DY]}{[TM] \cdot [SP]}$$

V) Gravitational time dilation within the super symmetry

The central position of any single event in relativistic wave mechanics leads to very complex processes, because of superposition and interaction of the involved forces, with limited freedom of the movements through space and time, and within bounds of collective system compulsions. Increasing mass aggregations oppose basic time, with the impact of increasing slow down of relative time development, staying in the present of relatively accelerated time processes. If a light wave leaves a large mass, an observation shows the effect of a red shift. This effect describes a gravitational dilatation effect. t_{0D}-development on a larger mass-system has reached a more advanced stage than on the remote and smaller mass. The time ratio $\Delta t_{0D}/t_{FM}$ which is measured on the larger mass is higher than the time ratio $\Delta t_{0D}/t_{F0}$ on the remote object, taking equal Δt_{0D}-pace of a t_{0D}-clock, and observed from remote positions with a fixed reference interval t_{FX}. Leaving the large mass, the length of oscillation periods in departing objects get relatively longer and an object's time runs relatively faster.

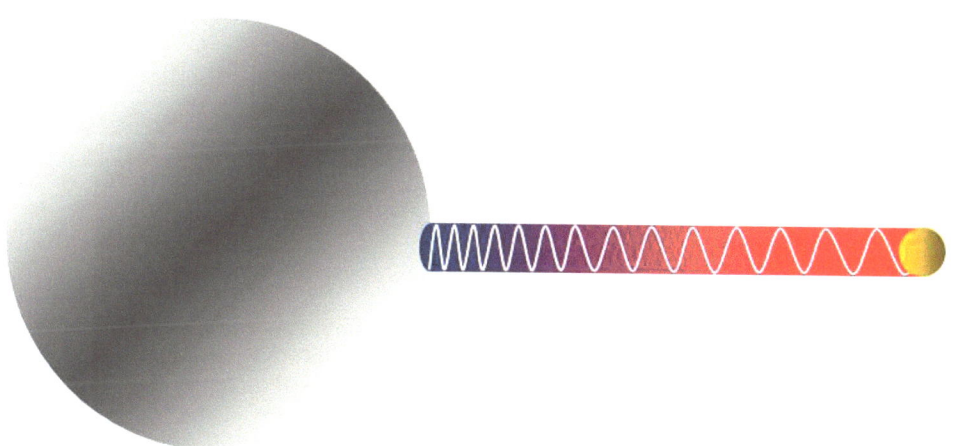

Figure 10: Gravitational red shift

If light is leaving from a large mass aggregation, it loses relative t_0/t_F-energy. Therefore, the light wave gets relatively longer. All objects leaving the gravitation field of a large mass decrease their t_{0D}/t_{F0}-ratio against the t_{0D}/t_{FM}- ratio of the large mass, due to increasing t_{FX}-resultants. This indicates the reduction of relative space-time tension.

Gravitational dilatation describes a picture of space-time embedded masses with respect to one common t_{0D}-time axis, with $v_{rel} = 0$, and the t_{0D}-starting time of measurement with $t = 0$. This measurement scenario is very similar to the relativistic projection of relative speed onto the t_{1D}-time axis, being a reference axis with $t = 0$ and $v = 0$. The measurement of the gravitational time dilation is related to t_0/t_F, with $v_{rel} = 0$ as the zero point of a relative measurement. Light shows the relative stretching of superimposed wave lengths if it leaves a mass in radial direction, and is losing energy in relation to its starting position. The t_0-axis shows that the clocks in a small object in space run definitely slightly faster. Subjective t_{0DO}-t_{1DO}-projections will stay perpendicular, being the reason for a constancy of speed of light.

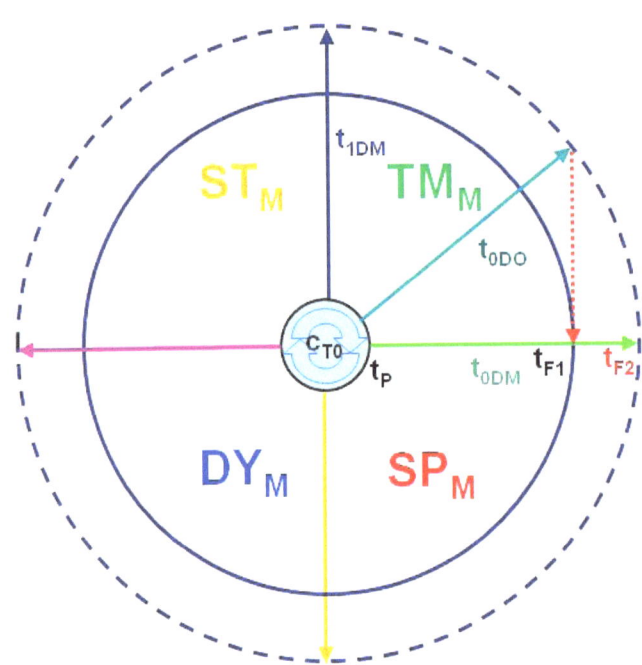

Figure 11: Gravitational dilation

Mass aggregations cause opposing space-time reactions, with individual values of t_F. Change of t_F cause the relative contraction or dilation of space-time frames, indicated by different circumferences. Relative velocity, however, rotates the t_{0DO}-vector of an object. Projection on t_{0DM} of the large mass (red arrow) shows the relative dilation of time in the moving object with respect to a measurement interval t_{F1}. The comparisons of individual masses are based on t_{0DM}/t_F-ratios, and on their corresponding space-time-frame differences. Therefore, the relativistic mass of any moving object increases with the t_{0DO}-dilation factor of the SRT.

Figure 11 combines time dilations by relative speed and by gravitation with equal impact but optically differentiated. Both could be described with rotations alone, as well. It shows how space-time tension is build up with these two parameters.

VI) Reflections upon electron mass and atomic mass unit

The following reflections upon electron mass and atomic mass units are of speculative character and based on the superposition of distinguishable E_Z-scenarios, each with 8 functionally different speed of light parameters and influences of trigonometry and geometry. A length-time-speed-acceleration diagram shows four basically equal perpendicular speeds of light axes and the E_Z-equation reveals scaffolding by four different $1/c$-statics parameters. Mass could be described by $\frac{X}{c^4}$-formulas: X describes the distortion strength of a basic construction with four perpendicular c-directions. Taking just X in the numerator requires a discussion of the impacts of the E_Z-trigonometry and of all eight speeds of light parameters of all involved superimposed E_Z-scenarios. Each E_Z-equation can describe the t_p-phase of events. Quantum mechanical numbers z_x determine the energy detachment, if particles are supposed to be released by a 180° turn of a minimum possible space-time area. There are eight c-parameters in E_Z that influence mass development in multifarious ways, if distinguishable E_Z-scenarios can be superimposed in multifarious ways, or in case of a distortion of one regular E_Z-scenario. One simple detached configuration that should appear with mass features could be described by the term $(2 \cdot 2 \cdot 2 \cdot 2)/c^4$, i.e. $2^4/c^4$, in case of a 180° turn with $z_x = 2$, linking four c-quantities: $2/c_{(DY)} \cdot 2/c_{(ST)} \cdot 2/c_{(TM)} \cdot 2/c_{(SP)}$ and separating them in space-time according to figure 4.

$\pm 16/c^4$-elements may be produced in pair as separated bubbles. They are assumed to form electrons and positrons, because they have the smallest possible value of a 180° detached E_{Z1}-configuration in the basic E_{Z0}-set-up with t_{0D}-time. The masses of electron-neutrinos and anti-electron-neutrinos appear with relative reduction because of their 90/270°-structures. The set-up with t_{1D}-time would identify these two particles as electron and positron.

The 16th tidal quantum of the term $16/c^4$ is supposed to not contribute to the mass but to develop the half integer spin asymmetry of the electron with its magnetic momentum. The mathematical generation of electron-positron pairs could be realized by ±180°-counter rotation, or coupled rotations of opposing areas, +180° each. The result will be identical because of sign changes of opposing axes by inverted order of vector cross products.

The approach to calibrate this rest mass with $c = 299792458$ m/s and force $F = m_e \cdot a$ with imposed acceleration a could be:

$$m_e = \frac{F}{a} = \frac{15}{c^4} \cdot C_A, \quad \text{with} \quad C_A = 4.9054685 \cdot 10^2 \text{ kg} \cdot \text{m}^4/\text{s}^4$$

applying an electron's rest mass m_e with $m_e = 9.1093825 \cdot 10^{-31}$ kg, fixing the strength of $1/c^4$-distortion with $C_A = 4.9054685 \cdot 10^2$ kg · m^4/s^4.

Any mass defect is distributed in space-time by opposing passive speed of light and acquiring perpendicular active speed of light, releasing the energy equivalence E=mc^2 because of these two perpendicular c-components. Let us suppose to not rotate ±180° for in-pair productions of electrons and positrons or electron-neutrinos and anti-electron-neutrinos, but to rotate E_z-configurations until receiving other feasible long-term stable detachment solutions. Instead of using a factor 15 of the m_e-equation, the nuclear rest mass could be substantiated by factor $\pi/2 \cdot c^{1/2}$. $\pi/2$ could represent the scaling factor of one of the trigonometric arguments in case of $\cos(\frac{\pi}{2}x) \approx \frac{\pi}{2}x$ and factor $c^{1/2}$ could be explained by simultaneous reduction of one factor c of the denominator to $c^{1/2}$ with a factor $|c^{1/2}|$ in the numerator. This picture describes the suppression of one c^2-sector, developing its c^2-spreading into one coinciding c-line, requiring the factor $|c^{1/2}|$ to realize a seeming loss of one of eight c-dimensions. The result would be the nucleus basic unit $1u_C$ that appears with a basic construction of combined core elements which could force electrons as by suppression liberated particles into an orbit around this nucleus. A reduction of speed of light parameters by coinciding E_z-space-time-grid elements develops stable detached energy formations.

$$1u_C = \frac{\pi}{2} \cdot \frac{|c^{\frac{1}{2}}|}{c^4} \cdot C_A$$

c = 299792458 m/s leads to $1u_C = 1.6516876 \cdot 10^{-27}$ kg

The atomic mass unit 1u is based on the chemical element carbon 12 and with a value of $1u = 1.660539 \cdot 10^{-27}$ kg. Accordingly, a proportion between $1u_C$ and $1u$ can be derived: $1u = 1.005359 u_C$.

Protons develop the rest mass $m_P = 1.012674 \cdot u_C$,
Neutrons develop the rest mass $m_N = 1.014070 \cdot u_C$.

This X/c^4 rest mass development scheme gets feasible, if rotary symmetry, superposition and reductions can form all elements of the periodic system.

VII) Expedition through the standard model of physics

Descriptions of nature with physics are orientated towards and tuned with our human perception faculties. All measurements register defined physical quantities that capture qualities and characteristics of objects. An observed object may be a concrete object, but could be also a physical state, or a process. The conformity to all physical laws is expressed by mathematical descriptions and logical linking of all physical quantities and processes.

Quantities are always described by a value and a connected unit.

Physics defines seven basic types of quantities. These are length, time, mass, temperature, electric current strength, substance quantity, luminous intensity. Other types of quantities can be derived from these basic seven types either by formulations of natural laws or, for specified purposes, with products and quotients. An RTS is based on a fundamental quantity only, tidal energy, with the system's acceptance of a relative change of flow and tensions that is perceived either as space lengths, as acceleration forces, or in various forms of condensing matter. All other types of quantities and all natural laws can be derived from a rip into tidal flows and tension energies.

An S-TQ-model of rotating interacting relativistic relative time developments opens complementary and alternative perspectives on all kinds of energy manifestations in nature. The descriptions of these processes extend the ways of looking into the details of physics, which is quite helpful to bring all natural phenomena into an overall context.

In the course of this chapter, I shall skim through the disciplines of physics, making cross references to compatibility with the RTS. These six disciplines are mechanics, thermodynamics, acoustics, optics, electrics, and nuclear physics.

Mechanics

Let us start with mechanics. The basic length to measure distances is today one meter [m]. It is the length, which light in a vacuum covers in the time interval of 1/299 792 458 seconds. Length of a relativistic time suite can be considered as a manifestation of subjective reactive power patchwork, and becomes a derivative quantity. The interval of one second [s] is a defined quantity, based on the radiation of cesium 133.

The mass unit "kilogram [kg]" is a defined quantity, and related to a regular mass of a kilogram reference prototype in Paris, made of platinum-iridium. The cylindrical reference mass has a height of 39 mm and a diameter of 39 mm. Mass is a relative quantity that is supposed to be constructed by c^{-4}-elements. It is relatively growing with increasing relative speed, proportional to time dilatation. In case of superimposed three-dimensional t_{3D}-vibrations caused by extreme temperatures matter is split up into its components. This status is called plasma, representing 99% of all visible matter.

The difficult part of re-thinking needs to be invested in the basic discipline of physics, namely kinematics. Two factors have to be considered carefully. On one side the movement on defined, only theoretically straight lines, if acceleration and reacting forces have to be described with application of integral calculus and with the quadratic speed terms. On the other side, it concerns the use of differential calculus, formally eliminating dimensions in case of using straight lines. Space-time-speed-acceleration diagrams reveal that every process should be calculated with the rotary space-time set-up.

Integral calculi of kinematics recover reductions of dimensions, in case of using straight motion lines in LWH. Therefore, all processes at low relative speed can be usually captured with sufficiently high precision.

Because of the relativistic three-dimensional spreading of t_{3D}-directions to all points of simultaneity and respective t_{0D}-spots, all length developments in LWH are straight throughout the visible universe into all radial outbound directions from any observation point, with an increasing retrospective view but visible light path deviations in the near of planets and suns, i.e. heavy mass aggregations.

Comments to Newton's three, non-relativistic axioms:

Axiom 1:
"Without external forces, any mass remains in the state of rest, or in the state of straight and steady motion". This characteristic of masses is called inertia. There are always gravitational forces involved, but with low impacts across enough long distances. The t_{1D}-tension factor is not considered in this non-relativistic picture of inertia, and thus no relativistic mass increase. The reason for the validity of this axiom for non-relativistic descriptions is the reactive power construction of space-time and a subjective evaluation of equally behaving space-time effects in any closed inertial system.

Axiom 2:
"Effective forces from initiated accelerations are always proportional to each other: $F \sim a$". Looking at the dynamic super symmetry, this fact becomes obvious. Each aligned force on a mass causes the shift of balanced forces between t_{0D} and t_{2D}, which has to be transcribed with a superimposed t_{2D}-acceleration, holding against relative t_{0D}-developments.

Axiom 3:
"If a mass body has a force impact on another body, it experiences from the other body an opposing force with equal strength". The reason for this is the central space-time positioning of each single event with respect to involved c_{T0}-compensation-points, causing all action-reaction processes of individual S-TQ-space-time-mass interaction scenarios above quantum thresholds.

Nevertheless, these three axioms of Newton capture only a small part of the processes within the entire system: they cannot explain influences of tidal vortexes and elementary angular momentum, and the effects of collective alignments. These axioms describe t_{0D}-t_{1D}-t_{2D}-t_{3D}-interactions at low speed. Relativistic mechanics allows time corrections in high speed scenarios with the formulas of the SRT in low gravity environments. However, any clock on earth runs approximately by factor 6.95317×10^{-10} slower than in far space.

The proportion between the mass of a defined body and its volume is called density. Cohesion and adhesion forces, causing the density of a body, react with increased t_{3D}-vibration in case of increasing the temperature, or in case of compression. Gaseous materials show up with temperature and pressure depending densities. In order to enclose a gas, there is the necessity of an opposing pressure that could be realized by appropriate materials with a sufficiently higher t_{3D}-compression or by interplay of cooling and gravitation, just like in the earth's atmosphere.

The springiness of mechanical springs demonstrates an elasticity of matter manifestations. The construction of elastic matter requires special material structures, allowing the stretching of material t_{3D}-distance with simultaneous t_{3D}-compression of other areas. This process initiates an increasing t_{3D}-vibration, causing the partial increase of material temperatures.

Within gravity fields of large masses, it is only possible to move with at least one component, opposing this force field. Aerodynamic locomotion, rocket propulsion and buoyancy of hot-air balloons are different examples of such opposing forces. But they all have one feature in common: their individual t_{3D}-structure of substance density supports lift, buoyancy, or propulsion in favor of gaining or maintaining relative t_{3D}-distance. All processes of motion are connected with an increased t_{3D}-vibration of parts and materials that are involved in the repulsion techniques to gain a relative distance.

Any form of energy in the universe is in principle kinetic energy. Because of the S-TQ reactive power construction and via residual angular momentum, there is the possibility to store kinetic energies in different forms of potential energies, which can be released in case of a necessity. Release of compact matter manifested energy is possible, but with an involvement of extremely high distortion forces, usually triggered by implosions or collisions.

Chemical states of substances are mainly a function of the temperature, in combination with various affinities to link and interact with other substances. Between molecules there are interdependent acting forces, responsible for chemical states and in case of solids and liquids responsible for volumes. Super symmetry adjusts an equilibrium distance between molecules.

The range of cohesion forces between the molecules of a body is low and the repulsive force decreases faster than the attracting force. The resulting sphere of interaction has the radius of approximately 10 nanometers [nm].

The cohesion of molecules of a substance and an adhesion of molecules of different substances can be designed and controlled by nanotechnology, in order to interact in more efficient ways. All products and processes can be tailored to applications and many innovative inventions are about to come.

Mechanical oscillation is a periodical spatial shift of masses, with an overall electromagnetic equilibrium in both transit spaces. All initiated t_{1D}-processes are captured by speed levels, t_{3D}-processes by length measurement, as two quite different manifestations of tidal energies.

Thermodynamics

Thermodynamics differentiates between thermal state and thermal energy. Thermal state takes temperature as summarized description of microcosmic t_{1D}-t_{3D}-vibrations of all elementary particles inside the atoms, of combined atoms or molecules, either free in space, or fixed by grid constructions.

For thermal energy, a principle of energy conservation is valid, too. Thermal energy can be compared with law conformities of mechanics, because it is merely motion energy of elementary particles, of atoms, and of molecules. Aggregation of matter reduces possibilities of three-dimensional movement inside and outside of gravity fields.

In return for these restrictions, it is possible to initiate motions in any radial direction of three-dimensional space, if there is any physical possibility or a technical solution to push off from other masses, initiating propulsions in opposing directions. If a mass becomes extremely large in relation to other masses, the large mass absorbs reactive forces of pushing, like our earth does.

Acoustics

Sound generation and dispersion are superimposed forces, generated by low frequency oscillations and waves, and distributed by suitable carriers. Applying only low amplitudes of sound oscillations shows measurement results of a dispersion speed that mainly depends on mechanical properties of the carrier, not on the frequency of the sound wave.

Any relative speed between a transmitter and the receiver causes the wave distortions, known as the Doppler-effect. Increasing constantly the distance from the receiver with respect to the observation at the receiver, initiates the relative stretching of the sound frequency of the transmitter, resulting in a lower frequency observation at the receiver. If, on the contrary, a transmitter approaches the receiver, the waves appear relatively compressed with the effect of the relative increase of frequency. The Doppler-effect differs from relativistic movements, because the speed of tidal energy developments is always speed of light that can relatively fall only below this original value in case of relative speed or time dilation by gravity. Therefore, SRT-dilatation is observed with any heading in space. The Doppler-effect, however, is the result of a non-relativistic translation of interacting t_{1D}- t_{3D}- points in space.

Optics

Light shows combined ST and DY-features, being an electromagnetic wave and a SP-TM-particle stream without rest mass, depending on the strength of an excitation. Light follows the t_{1D}-axis in relation to the t_{0D}-course of time.

The interactions in atomic grid structures of a medium reduce the speed of light, seemingly. This effect is triggered by microcosmic processes, if light is sent through any medium. Reduction of relative speed is combined with relative rotation from t_{1D} towards t_{0D} and from t_{3D} towards t_{2D}. t_{2D} causes the refraction of a slanting ray into the direction of the denser medium towards the plumb, because a t_{0D}-development points to the direction of the denser medium. In case of sufficiently flat slanting angles, t_{0D}-deflection develops a total reflection, if the light ray passes from the denser medium to the thinner medium. This way, the ray will stay inside the medium, like in fiber glass wires.

Techniques of optics bundle light rays with convex lenses or disperse ray with concave lenses using material deviation capabilities by slowing down.

A comparable scenario to refraction of light is the diffraction of light with deflection into t_{0D} interactions lines between the photons and the diffraction edge. Observations from the viewpoint of the light ray interpret the t_{0D} time course as a length. This means that the way for a light ray becomes longer. An observation of t_{0D} in a t_{0D}-position interprets the t_{0D}-course as pure time development. This means that the light ray needs just more time to pass the denser medium and is in fact slower because the length stays unchanged with non-relativistic t_{0D}-perception.

Polarization of light is possible because transit subspaces are relativistic flat constructions in a three-dimensional space and photons are provided with the characteristics of these flat spaces. Any wave is linearly polarized if it oscillates in one direction with respect to the direction of dispersion.

Electrics

Electricity is a description of space segment ST and a coupling with DY and SP, combined with the passive speed of light function of TM for ST. Motion of electrons generate electric current and rotating magnetic fields. This kind of excitation of interactions between ST and DY and thus also between SP and TM is bound to an existence of SP-matter. Electric current experiences electrical resistance in materials, causing molecular t_{3D}-vibrations that result in increasing material temperatures in comparison with the environment. A possibility to reduce electrical resistance of electrical wires is cooling down close to the absolute zero level T_0. The result is the superconductivity effect with very low resistance because t_{3D}-vibrations are figuratively frozen if the temperature gets close to T_0 and electrons experience only low resistances across middle way lengths. Specialized high-tech laboratories are working on alternative nanotechnological structures, providing this superconductivity effect already under environmental conditions of earth's surface.

Because of the transverse tidal constructions, electric field lines are always leaving perpendicularly to the surfaces of conductors, and, according to the definition, disperse from positive poles towards negative poles. Movements of electrons within an aligned electric field are perpendicular to a generated rotating magnetic field. Electromagnetic interactions can be synchronized throughout the spreading range of electromagnetic waves. Without angular tidal momentum any synchronization like in antennas would be impossible.

Atomic and Nuclear Physics

Atomic and nuclear physics describe various forms of elementary particles and formations and interactions within atomic structures. It is the description of tidal flow and tension around c_{T0}-spheres causing space-time changes and nuclear counter forces.

Dimensionally reduced pictures of atom constructions cause the necessity for constants of nature and postulated energy relations. S-TQ simplifies the comprehension and assessments of energetic set-ups and motions within nuclei and atomic shells, because it broadens the equilibrium constellation view of the TM-SP-perspective, i.e. it portrays processes in time and space in a way, any neutral observer would perceive it in a monitoring position that is completely outside of the entire system. Highly energized particles are expected to be unstable E_Z-constructions (myons, etc.).

Artificial generation of entire anti-particle-structures of atoms in SP requires special conservation techniques to keep them in the wrong opposite space-time-speed-acceleration segment. Research centers for the production of anti-matter that are developing technical solutions for a proper conservation with an artificial t_{2D}-t_{1D}-time-space-environment are in the planning stage.

Literature references:

1) Special theory of relativity:
Albert Einstein, *Zur Elektrodynamik bewegter Körper, Annalen der Physik und Chemie.* 17, 1905, S. 891–921

2) General theory of relativity:
Albert Einstein, *Annalen der Physik*, 49, 1916, S. 769-822

3) Constancy of speed of light:
First experiment by Michelson-Morley in 1887
http://www.relativitycalculator.com/Albert_Michelson_Part_I.shtml

4) A Treatise on Electricity and Magnetism Vol. 1 and 2
James Clerk Maxwell, 1904-edition with corrections – Antique Books Collection

5) Raum und Zeit
Hermann Minkowski, lecture at the 80th Naturalist Assembly in Cologne, Germany on the 21st September 1908

6) New significance for the cosmological constant of the general theory of relativity
http://hubblesite.org/newscenter/archive/releases/2009/08/

7) Dark energy in astrophysics
The Cosmic Triangle: Revealing the State of the Universe

Neta A. Bahcall, Jeremiah P. Ostriker, Saul Perlmutter, Paul J. Steinhardt

Science 28 May 1999: Vol. 284. no. 5419, pp. 1481 - 1488 DOI: 10.1126/science.284.5419.1481

8) Uncertainty principle
W. Heisenberg (1930), *Physikalische Prinzipien der Quantentheorie* (Leipzig: Hirzel). English translation: *The Physical Principles of Quantum Theory* (Chicago: University of Chicago Press, 1930).

9) Planck time in quantum physics:
Max Planck, *Über irreversible Strahlungsvorgänge, Sitzungsberichte der Königlich Preußischen Akademie der Wissenschaften zu Berlin* 5, 1899, S. 478-80